The Figure Skate

The Figure Skate

A Research into the Form of Blade Best Adapted to Curvilinear Skating

By H. E. Vandervell

Edited by B. A. Thurber

Skating History Press

Main text originally published in 1901.
Introduction and notes © 2020 B. A. Thurber.
All rights reserved.

ISBN: 978-1-948100-06-9
LCCN: 2019920935

Skating History Press
Evanston, IL
http://www.skatinghistorypress.com/

Contents

Introduction 1

The father of English skating 3

Technical aspects of figure skating 9

This edition 13

The Figure Skate 15

Introduction 19
    Then and now: A short sketch 19

Chapter I. The blade 27
    Height of the blade from the ice 28
    The blade as a cutting instrument 28
    The thickness of the blade 29
    Experiments with a skate having only one edge 30
    The radius of the blade 32
    Skate blade with an adjustable radius 36

Chapter II. The radius of the blade in an inclined position 37
    Penetration or depth of the cut in of the blade 38
    The angular position and the other factors employed in their construction 39

Explanation of Plate I 39
Explanation of Plate II 39
Preliminary investigations and experiments . . 41
Explanation of Table 1 43
Explanation of Table 2: The angular position 44

Chapter III. Skate blades whose sides are non-parallel   **49**
The convex sided blade 51
The concave sided blade 52
Convex and concave blades 55

Chapter IV. Measurements   **65**

A digression—Thoughts on the possibility of describing the hypocycloid curve by means of turns   **73**
The epicycloid 76
The cycloid 76
The hypocycloid 76
The paradox 77

Commentary   79

Notes   81

Further reading   101

Bibliography   103

Illustration credits   105

List of Figures

1	Vandervell's first skate.	21
2	Vandervell's first design.	22
3	Rodgers' patent spring figure skate	24
4	Acute, right, and obtuse blade angles.	29
5	A double radius.	35
6	Plate I	40
7	Plate II: Radii	42
8	Convex and concave blades.	50
9	Side view	50
10	Plate III: Models of curves.	58
11	Plate IV	60
12	Plate V	61
13	A circle and an ellipse.	62
14	Plate VI	74
C1	Filor's improved skates	86
C2	The Krause skate	87
C3	The geometry for the calculations in Table 1.	88
C4	The geometry for the calculations in Table 2.	89
C5	The Dowler skates in the editor's collection.	90
C6	Acute, obtuse, and right edges.	91
C7	A cycloid.	94
C8	A curtate cycloid.	94
C9	A prolate cycloid.	95

Introduction

The father of English skating

Henry Eugene Vandervell (1824–1908) has been called the "father of English style skating." His achievements in skating are well-known to historians of the sport. They include inventing the counter turn, writing (with T. Maxwell Witham) *A System of Figure Skating*, and chairing the Ice Figure Committee of the National Skating Association.[1] Among skaters, little is known about Vandervell's personal life, and despite his stature, no biography has been written. Using online genealogical databases, I've made a start at uncovering his life. Here is a general outline. Certain events, noted below, are ripe for further exploration.

Vandervell was born to Francis Vandervell, a shoemaker, and his wife Mary and baptized on December 17, 1824 at St. George Hanover Square in Westminster.[2] In 1851, the census records 26-year-old Henry Eugene Vandervell working as a stockbroker and living in his mother's house at 82 St. James's Road with his sisters Fanny and Mary, Fanny's husband John Camp-

---

[1] James R. Hines, "Vandervell, Henry Eugene (1824–1908)," in *Historical Dictionary of Figure Skating* (Plymouth, UK: Scarecrow Press, 2011), 233.

[2] "Board of Guardian Records and Church of England Parish Registers," digital image s.v. "Henry Cugene Vandervell," *Ancestry.com*.

bell, Mary's daughter Mary P. Lewis, two house servants, a cook, and a groom. They also housed a visitor, Robert Rose, on census night.[3]

In 1854, Vandervell married Rebecca Batt, born to Samuel and Jane Batt and baptized on October 25, 1823.[4] They moved into their own place in Kensington, and the 1861 census records them living together with a single servant and their twelve-year-old daughter, Lavinia.[5] Lavinia Jane Batt was born to Rebecca Batt and christened on April 25, 1849.[6] No father is listed in the record, and Vandervell and Rebecca did not marry until May 4, 1854, when Lavinia was five years old.[7] Regardless of whether Vandervell was her biological father, he played that role in her life. He gave her away at her wedding to Montague Weatherly Marriott on January 24, 1871.[8] In his will, Vandervell recognized Lavinia as his daughter and left her a small inheritance:

> I give and bequeath to my daughter (by

---

[3] 1851 England Census, Paddington, London, digital image s.v. "Henry E Vandervell," *Ancestry.com*.

[4] "Surrey, England, Church of England Baptisms, 1813–1917," digital image s.v. "Rebecca Batt," *Ancestry.com*.

[5] 1861 England Census, Paddington, St. John, Middlesex, digital image s.v. "Henry Eugene Vandewall," *Ancestry.com*.

[6] "England Births and Christenings, 1538–1975," s.v. "Rebecca Batt," *FamilySearch.org*.

[7] "England Marriages, 1538–1973," digital image s.v. "Henry Eugene Vandervell and Rebecca Batt, 04 May 1854," *FamilySearch.org*.

[8] "London, England, Church of England Marriages and Banns," digital image s.v. "Lavinia Jane Vaudervell," *Ancestry.com*.

my former wife) Lavinia Jane Marriott of Telford House Sternhold Avenue Streatham the wife of Montagu Weatherly Marriott the sum of five hundred pounds for her absolute use and benefit in addition to the provision I made for her in her Marriage Settlement but to her exclusion from any other benefits under this will.[9]

Rebecca died in 1867,[10] leaving Vandervell a rich widower. During his marriage, he had done well in his career—well enough to afford a lease on the townhouse at 28 Aldridge Road Villas, currently worth about 4.5 million pounds, according to zoopla.co.uk. After Rebecca's death, Vandervell's wealth doubtless made him highly desirable as a husband.

On March 12, 1869, he married Fanny Thornton, daughter of Joseph Thornton.[11] Her birth in 1851[12] made Fanny just 19 at her wedding, 26 years younger than Vandervell, and two years younger than Lavinia. Did she skate? Given Vandervell's involvement in the sport, I would expect so. His (and T. Maxwell Witham's) remarks on ladies skating in the first edition of *A Sys-*

---

[9] Henry Eugene Vandervell, will dated March 14, 1984, proved October 6, 1908, Principal Probate Registry, UK. Copy in possession of editor.

[10] "England and Wales Death Registration Index 1837-2007," digital image s.v. "Rebecca Vandervell, 1867," *FamilySearch.org*.

[11] "West Yorkshire, England, Church of England Marriages and Banns, 1813–1935," digital image s.v. "Fanny Thornton," *Ancestry.com*.

[12] 1881 England Census, Paddington, London, digital image s.v. "Fanny Vanderwall," *Ancestry.com*.

tem of Figure Skating—published the year before he married her—suggest that if she didn't already skate, he would have encouraged her:

> We can scarcely imagine a more delightful, exhilarating, and health-giving exercise for ladies in winter-time than skating... We rejoice to think that within the last few years the girls of England have been taking to skating in considerable numbers.[13]

Vandervell and Fanny had nine children, seven of whom were still living in 1911.[14] Their children proved a bit slippery to track down. The eldest, Henry Eugene, was born in 1870. Second was Charles Anthony in 1871. He went on to do important work on the electrical systems in cars.[15] Next came, in order, Percy, Arthur, Ethel, Maud, and Francis.[16] Those are the seven who were living with their parents in 1891. But there were nine altogether. The other two were hard to

---

[13] H. E. Vandervell and T. Maxwell Witham, *A System of Figure-Skating: Being the Theory and Practice of the Art as Developed in England, with a Glance at Its Origin and History* (London: Horace Cox, 1869), 230–231.

[14] 1911 England and Wales Census, Paddington, London, digital image s.v. "Fanny Vandervell," *FamilySearch.org*.

[15] See his entry in "Grace's Guide to British Industrial History" https://www.gracesguide.co.uk/Charles_Anthony_Vandervell.

[16] 1891 England Census, Paddington, London, digital image s.v. "Fanny Vanderwall," *Ancestry.com*.

find. One seems to be Flora, born in 1884,[17] who may not have survived more than a few years. The other may have died in infancy.

Whether and how well they skated is a topic for future research. Presumably, they were all fine figure skaters, like their father. Vandervell was also an accomplished cornet player who performed at the party the London Stock Exchange held for the Queen's eightieth birthday on May 24, 1899.

> Under the conductorship of Mr. Clarke two verses were sung, the cornets being of immense service in preserving the original key, and in supplying the intermediary cadenzas. The novelty was warmly applauded all round, and Mr. H. E. Vandervell, a venerable member of the house...was heartily complimented upon his playing.[18]

In his will, Vandervell mentions "musical instruments" among the goods left to Fanny, along with his lathe and mechanical tools. A codicil dated March 14, 1894, gives a bit more detail, specifying cornets and "sporting effects" (presumably including his skates).

When Vandervell died on September 13, 1908, he left his estate to his wife and five of his surviving children. It was valued at £35,145 14s. 2d.[19]—the equivalent of £4,153,092 in 2018, according to the Bank of

---

[17]"England and Wales Birth Registration Index, 1837–2008," s.v. "Flora Vandervell," *FamilySearch.org*.

[18]Charles Duguid, *The Story of the Stock Exchange: Its History and Position* (London: Grant Richards, 1901), 411–412.

[19]Henry Eugene Vandervell, will.

England's inflation calculator.[20] His first and third sons, Henry Eugene and Percy were named executors.[21] Henry Eugene, Charles Anthony, and Percy each received £1500 to invest. Daughters Ethel and Maud each received an annuity of £60 per year for as long as their mother remained a widow plus £100 for a wedding outfit (assuming their mother approved of the marriage!). When Fanny's status changed, they were to receive £2000 each as a lump sum.[22]

Remember that seven of nine children were living in 1911, according to the census record. If we suppose that Flora and the nameless child died in childhood, that leaves Arthur and Francis to account for. It seems that they were still alive when Vandervell died, but they are not mentioned in his will. The 1911 census lists an Arthur Vandervell born in London in 1874, just like our Arthur, living in Basingstoke and working as a fruit and poultry farmer.[23] Francis, born in 1886,[24] remains a mystery. I wonder what family drama led to them being left out of their father's will.

---

[20]https://www.bankofengland.co.uk/monetary-policy/inflation/inflation-calculator.

[21]Henry Eugene Vandervell, will.

[22]Henry Eugene Vandervell, will.

[23]1911 England Census, Basingstoke, Hampsire, digital image s.v. "Arthur Vandervell," *FamilySearch.org.*

[24]"England and Wales Birth Registration Index, 1837–2008," s.v. "Francis Vandervell," *FamilySearch.org.*

Technical aspects of figure skating

Vandervell's main legacy is in his contributions to figure skating. He was inducted into the World Figure Skating Hall of Fame in 2015.[25] *The Figure Skate*, an account of his experiments with different types of blade, was published in 1901, when he was 77 years old. Today's adult skaters should find this inspiring: Vandervell was able to skate well into old age. Of course, he may have done the experiments over a long period. He was still experimenting as late as 1890, when he was 66, according to his letter to *The Field* describing his experiments with a single-edged skate (reprinted on pages 30–31).

Vandervell's will mentions his lathe and machine tools among his effects. These underscore how serious he was about skating. This was not merely a pastime for him; he wanted to figure out how skates worked and improve them scientifically. The main variable he optimized in this book is what we now call the rocker radius: the curvature of the blade from front to back. The other curve on a modern figure skate is the radius of hollow: the curvature from edge to edge. The radius of hollow is easily changed. Skaters can ask for a specific hollow—usually between $\frac{3}{8}$ and $\frac{5}{8}$ of an inch now—when they get their skates sharpened. Vandervell preferred not to have a hollow at all; his skates were flat side-to-side.

---

[25] Visit the Hall of Fame at http://www.worldskatingmuseum.org/WorldHallOfFame.html

In contrast to the radius of hollow, the rocker radius is fixed. Vandervell tested skates with rockers between three and nine feet. He recommended a radius of six feet "for all-round purposes," but three feet for small figures, like loops, and nine feet for combined skating[26] (page 34). This is not far from the rockers used today. Today's figure skates come with a rocker radius of seven or eight feet that may vary along the length of the blade. This is right in between Vandervell's recommendations.

Vandervell also evaluated the effectiveness of changing the blade's thickness along its length. In chapter III, he concluded that keeping the sides of the blade parallel and the thickness constant is best. Most blades today are parallel, but some newer blades are tapered or "parabolic." It is not clear what effect this has on the skater's ability to skate. Vandervell argued that, at best, the effect is negligible, and it could even make things worse (page 50).

The methods Vandervell described in this book are scientific. He was very careful with his measurements and enlisted help with the mathematics when he needed it. The book is based on this sound science, and the recommendations are not far from today's preferences despite substantial changes in skating style. Back then, nobody was doing quadruple (or even double) jumps.

The book ends with a digression on skating rather than types of blade. Vandervell was unable to skate

---

[26]In combined skating, a group of skaters—most often four—skate prescribed movements in parallel.

what he calls a "hypocycloid curve." He left it as a challenge for his readers. I think it will not be so difficult for today's skaters.

This edition

This edition follows the text of Vandervell's originally work closely, with a few minor errors corrected. I reformatted tables 1 and 2 to make them fit the pages better and renumbered the figures. All the figures in the main text were present in the original; I have added captions to some figures in the main text and texts quoted in the commentary for the table at the beginning of this book, but only the original captions are included in the text. The footnotes in the text, designated by symbols, are Vandervell's. Additional commentary has been added in the endnotes, which are referred to by numbers in the main text.

The Figure Skate

Preface

In this little work, in addition to other matters connected with the skate blade, I have inserted some tables which I have worked out.

These give the length of the skate blade actually below the surface of the ice and in contact with it, under certain fixed conditions.

It is obvious that a knowledge of the effect produced by such contact is absolutely necessary before any reliable opinion can be formed as to the best radius to adopt.

The subject has required numerous experiments, besides one hundred and seventy one geometrical and trigonometrical calculations.

Ten of these, relating to the convex and concave forms, are very difficult and more in the way of the mathematical expert.

I am indebted to Professor John Graham, B. A., B. E., Demonstrator and Lecturer on Applied Mathematics at the Technical College, London, for kindly working them out for me.

For the remaining one hundred and sixty one calculations I am responsible.

A digression from my main subject enables me to give in a short final chapter my experience of a geometric figure which I formerly tried to resolve in vain, and therefore called it The Paradox. It is the hypocycloid curve, and is to be attempted by turns.

H. E. VANDERVELL

28, ALDRIDGE ROAD VILLAS,
WEST BOURNE PARK, W.
February, 1901.

Introduction

Then and now: A short sketch

The figure skate blade with parallel sides, and with its base curved instead of straight, was undoubtedly derived from the travelling or speed skate, but at what period is unknown, probably in the beginning of the seventeenth century.

Well then, looking back into the far off, the name of the enthusiastic individual who first ground his skate irons to a curve is wanting, but the wish to turn round more easily in small circles, and eventually to practise curvilinear skating, must have been the incentive.

The adaptation was probably effected by grinding off a little from the toe and heel part of the base of the iron of the speed skate, and thus forming an arc or curve, technically called by modern figure skaters, the radius.

Rough and imperfect doubtless in the first instance, yet to a certain extent effective for the purpose intended, it may safely be assumed that the man who first started skating in curves had but a rudimentary knowledge of the perfection of the principle involved in his apparently simple ice engraving instrument. He was in all probability unaware that he was entering the borderland of a beautiful art.

The progress of modern figure skating during the Victorian era has been rapid.

Curves of great difficulty have been conquered whilst

in the balanced position and have been engraved upon the ice, and many movements have been added to the combined figures.

It is not to be supposed that figure skaters who have accomplished this satisfactory work, were content with the comparatively crude instrument of a former period; over and over again, time after time, have men sought to improve it.

As difficulties in the art began to appear, skaters evidently thought that these might be more easily surmounted with fanciful alterations of the skate, beyond what were really sterling improvements.

Having myself taken a very modest part in these, as an amateur mechanic with skating propensities, my long experience may be of some little interest, for I happen to be one of those who have been in touch with the gradual employment of a much better skate.

It is well to look back sometimes; it gives a better impression of the obstacles to progress which had to be encountered in the early days of figure skating.

To begin a very short retrospect—the figure skate of the past was but little removed from the speed skate. It is represented in Fig. 1. Such was the skate I used in 1837. There were several old pairs in the country house in which I resided, so evidently it was the recognized form then, and it was the pattern with which I learnt forward skating.

Personally, however, when I commenced back skating, I found to my cost that these skates were very

The Figure Skate

dangerous. Ice was never cared for as it is now—stones, sticks, grit, chips and leaves were all too numerous on the usual places of resort.

This particular skate (Fig. 1) with the heel of the foot but partially supported, the blade ground out at the back end to a sharp angle, could not in many instances glide over a slight impediment. Falls consequently were numerous, and often of great severity, to prevent which the skater had to lean very much forward, and the progress in the art was utterly checked.

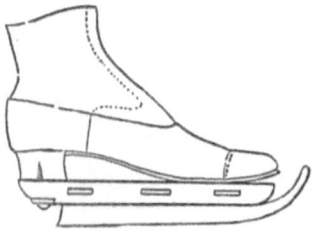

Figure 1

Then the thought naturally occurred to me: Why should not the blade of the skate be extended to the heel and rounded off?

To test this idea I made in the year 1840 a full sized model of a skate of this character, with a foot plate of metal, which I fancied a skate should have, and took it to a clever whitesmith in the town of Croydon, and after many consultations it was turned out thus:—

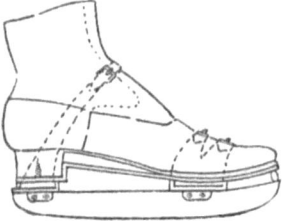

Figure 2

The method of fixing the blade to the foot plate was by two brass plates having longitudinal slots to hold the blade, and fitted to it with screws, such plates being riveted to the foot stock. The said brass plates had at the corners very short basses or legs, affording a space for the straps to pass through.

The elongation and rounding of the heel end was happily an improvement that came to stay. I bring it forward now from the fact that it evidently marked an epoch and was a forecast. Remove the straps, screw the iron foot plate to the sole of a boot, and we are on the way to the modern skate.

In those early days to which I refer, this skate made a little stir amongst my skating friends, some of whom were students at the far famed Addiscombe College, near to which I resided. Generally the innovation of the extended round-backed end was coldly received, but by a few others it was highly thought of. These skates were made entirely for my own use, but those who wanted similar ones were referred to the maker; lads of sixteen care not for patents.

I used these skates for several years, and as they

had the remarkably small radius of two feet, turns were easily acquired, but the skating became too small and contracted for the larger movements required in combined figures. I therefore discarded them in favour of blades with a much greater radius.

But although my skate was an original invention, as I have detailed, a prior claim to the elongation and rounding of the back end undoubtedly belongs to a Mr. Boswell.

In the year 1836, four years previous to my idea, this gentleman, who resided at Oxford, had some skates made after his pattern with elongated and rounded ends, and these were used in the Oxford Club (*vide* "A System of Skating," T. M. Witham.)[1]

It is therefore certain that in his locality Mr. Boswell was instrumental in making known the use of his improved skate, and that four years later, although quite unacquainted with him, I must have worked on somewhat similar lines with my skate.

As far as is known at present, there does not appear to have been anyone prior to Mr. Boswell connected with this improvement.

No doubt the innovation was a long time in abeyance, but at last it commenced to "catch on," and blades with the prolonged and rounded ends began to be generally known as the Club Skate, undoubtedly from this very Oxford Club alluded to. When I joined the Skating Club[2] in 1855 some few members continued to use square heels.

About the time I was making my experiment, or soon after, Rodgers' Patent Spring Figure Skate was brought out with some éclass. The foot plate was very ingenious and novel in action, and was rarely understood. I tried it occasionally, but the form of the blade was little removed from the antique—prolonged at the toe, square at the heel, and not much extended, and very slightly curved; in fact, to a very large radius.

The novelty consisted in the spring, and a drawing of this skate is given on page 24.[3]

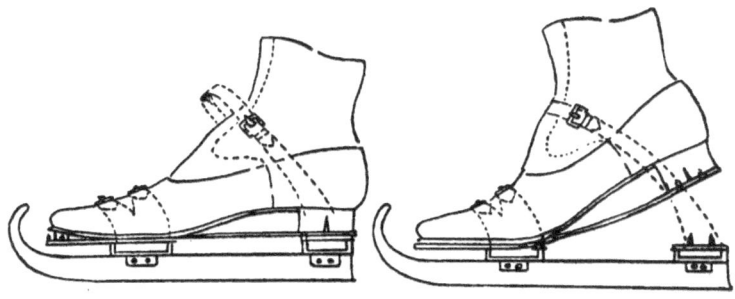

Figure 3

The idea was that the skate blade should always remain on the ice on the stroke. The play of the spring was the equivalent of the action of the skater, when lifting the heel of his ordinary skate to strike from the toe. There was necessarily a very loose strap at the heel over the instep, only tightened sufficiently when the skate was open to prevent the peg coming out. The two little points that may be noticed at the heel went through holes in the spring foot plate, so that when this was down these little pegs entered the heel of the boot and gave the proper stability.

This skite soon went out of fashion. The safety of the skate was dependent on the spring. There was a terrible side strain upon it in certain positions, springs broke and skaters were injured, and the skate was only used when the heel strap was tightened up, to bring the spring foot plate right down permanently, so that the sole use of the spring then was to keep the peg in the heel.

A friend of mine purposely broke off the spring and used the skate safely without it. I have noticed this skate because of the ingenious idea, a spring foot plate, being quite unique.

In a mechanical age like the past and present it was to be expected that something would be done to do away with the troublesome method of fastening skates on with straps.

The revolving tee plate with clamps was brought out with a fair measure of success. But the advent of the Acme and Barney and Berry, with adjustable metal fastenings, have had a well-earned and extensive popularity.

Skates, however, with a wooden bed, screw and straps, are still very extensively used.

After a long experience, the best figure skaters of the present day use the Mount Charles[4] fastening, in which the skate is permanently screwed to boots set aside for the purpose. But whatever may be the mounting of the skate, as long as it is comfortable and safe, is of quite secondary importance compared with the form of the blade, and as much divergence of opinion obtains as to this, it is well to make a thorough inquiry as to the reason of it.

This I have attempted to do in the following pages.*[5]

---

*NOTE.—To those who take an interest in the gradual evolution of skates generally from the earliest times I commend a most interesting, concise and useful little work entitled, "On the Outside Edge," written by my friend, Dr. G. H. Fowler, which will enable the reader thereof to get a good grip of this part of the subject.

Chapter I. The blade

The material of which the blade is made is a combination of iron and steel, and should be of the very best quality, rolled or welded together, the hard steel being at the base. The object of this is that the iron gives strength and toughness and supports the steel, which has to be hard tempered.

In learning, the skate blade is often unwittingly dashed down upon the ice with very great force, and nothing but this combination appears to give the requisite safety.

Skate blades have been made all of steel, but necessarily are not in favour.

Whether any alloy will supersede the combination of iron and steel is doubtful.

The base of the blade should be a part of a true circle, it should project $\frac{1}{8}$-inch in front and $\frac{1}{2}$-inch behind, the corners should just be rounded off.

In my opinion it should have parallel sides and be only $\frac{1}{8}$-inch thick; the reason for these conclusions I hope to demonstrate. The skate makers think that $\frac{1}{8}$-inch is hardly strong enough; I never had a fracture of this thickness in upwards of 20 years' experience. The blade can be thickened if thought necessary, commencing $\frac{1}{8}$-inch above the base, which will give it ample strength.[6]

It should be fastened to the frame by side screws and should thus be interchangeable with other blades.[7]

The whole being secured by its foot plate to the sole

of a close fitting boot, on the now well-known Mount Charles System. As to the employment of the blade in the stroke, it has been set forth in books on skating, and except that the angle at such a time is very much increased, involves no scientific interest.

Height of the blade from the ice

The distance the skater is elevated from the ice is generally $1\frac{3}{4}$ inches, but in some few cases, notably in the skates used by foreign skaters, this is much exceeded. Skates appear higher now than they used to when the foot stock was of wood. The lower the skate is, as long as it clears the ice well on a lay over, the better. I think there is a tendency unduly to raise the skate, but still if the man is below the middle height, and has strong ankles, it may be of advantage to be well above the ice, as it gives him a little more power.

The blade as a cutting instrument

Seeing that one of the offices of the skate blade is to cut a shallow curved groove in the ice, for the purpose of forming a base of resistance to the angular position of the skater, we have to consider the transverse section best adapted for the purpose. Three forms have been and are employed; the first is the acute, the second the right angle and the third the obtuse angle.

Fig. 4 (left) has a tendency to dig in, not only on account of the acute angle, but from the facts that the base is wider than above it and the side contact and

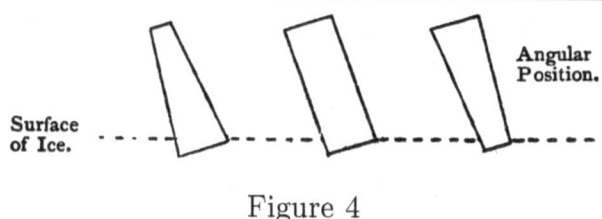

Figure 4

pressure are relieved. Fig. 4 (right) on the contrary, owing to the obtuse angle, is not so sharp and does not dig in at all. The base, however, being smaller than above it, there is a tendency to thrust away from the groove, and more side pressure. Fig. 4 (center) is a right angle, the cut and pressure and contact appear equally distributed, and I therefore think that it is the best form.

A blade that is too sharp, as we sometimes find when fresh ground, is an abomination.

It is quite astonishing what a power it exercises over the skater, who really then is not his own master. But the arbitrary curve is beautifully cut.

The sharpness soon wears off and can be remedied at once by drawing the edge very carefully and slightly over an oilstone.

The cut in is necessarily much affected by the radius.

The thickness of the blade

This is a factor in the blade of considerable importance in relation to the angular position of the skater. The flat of the blade, or that part of its surface which slides over the ice, forms with the side a right angle. For the purpose of obtaining a base of resistance to unnecessary penetration, it is not at all imperative that the blade should be thick. In practice it has been hitherto very variable, viz. from $\frac{7}{16}$-inch to $\frac{1}{8}$-inch. Personally I have for a considerable time entertained the opinion that the more the flat part of the base could be reduced in thickness the better, and that it would be well if possible to use a V shape, having therefore only one edge, and it was for the purpose or testing this that I made the following experiment, as detailed in my letter to *The Field*, which I quote from.

Experiments with a skate having only one edge

Sir,—Three years since, wishing to experiment with a skate having only one edge, I had a pair made; irons $\frac{1}{4}$-inch thick, ground to a knife, or V, edge; radius 6 ft. First trial on good ice in a state of slow thaw, V skate on right foot, ordinary skate on left; V skate cut in far too much, and wedged out great pieces of ice; ordinary skate perfect.

Second trial, in the succeeding winter, on ice at three degrees of frost,[8] the irons having been re-ground to an angle of 45°: V skate did not cut in too much but was very difficult to control and to balance upon.

Third trial much the same, but I managed to skate a serpentine line. This was very interesting, as only one edge existed to be employed in the operation; consequently, there was no distinction between the inside and outside except the lean over.

To get a good stroke with such skates is impossible, as at the necessary angle the bevel touches the ice and the edge slips. I attempted a few turns, without success; no encouragement was offered by the implement, and a sense of insecurity was always present, which vanished immediately the left foot with the ordinary skate was used.

H. E. VANDERVELL.

2A, Copthall-court, *Feb. 15th*, 1893.

The ordinary skate mentioned in this trial was only $\frac{1}{8}$-in. thick.

Theoretically, the V form of transverse section ought to be perfect, as giving a central thrust, and therefore reducing the strain on the ankle; but practically I consider it a failure, for in addition to other defects it does not relieve sufficiently in skating turns. For the skate with a flat base, be it ever so narrow, is admirably adapted to facilitate turns, because it helps very much to lift the blade out of the groove, which the V form does not.

Now, it being an axiom in figure skating that the skater inclining sideways should be on his edge as soon as possible, it follows that a narrow iron effects this at less inclination than a broad one. Consequently, it is much safer, and well suits the incessant changes of edge required in modern figure skating. For the very large curves used in the combined figures it is the only means of feeling the edge when the angle is perhaps only 5°.

In corroboration of this, I have heard rink professors remark that many of our combined figure skaters are often on the flat of the blade, and run the risk of being unsteady. This may be the case with broad irons, but not with narrow ones.

Moreover, the difficulty of teaching the outside edge is to get the learner off the flat of the blade; therefore, again, the narrower it is the better. As to my own experience, I have used a 9 feet radius of $\frac{1}{8}$-inch thick with great satisfaction for many years on hard, and also on soft, ice. I find that $\frac{1}{8}$-inch is ample to check the cut in, and form a sliding base, and it must be observed that $\frac{1}{8}$-inch is really only $\frac{1}{16}$-inch from centrality, and is as near the desired perfection as practicable.

The radius of the blade

This affects the skater's action in the highest degree and most sensitive manner. Apparently so simple, it is full of subtlety. Does the theory of skating point to any given radius as superior to another?

The answer to this question is not at first apparent, for it is dependent upon another question equally abstruse, viz.: How much of the curved blade ought to ride in the ice, allowing for the penetration or cut in, in order that the skater not only should have facility in making turns, etc., on the epicycloid curve, but proper stability on the large sweeps mainly of the character of concentric curves?

Now, before these two questions can be answered there must be added to them the consideration of the variable depth of the cut, according to the density of the ice, the sharpness of the skate, the weight of the skater, and the centrifugal force employed.

We are thus face to face with a very difficult problem indeed, impossible correctly to solve theoretically, as the basis for the calculation of one given radius only, as applicable to all kinds of figures, is wanting. Fortunately it is here that practical experience comes to our assistance with two well-established facts.

It is certain that a blade of 2 feet radius sets up automatically more than enough of turning power. A 9 feet radius gives all power of keeping out on big curves, and if it is assisted by imparted rotation, even loops can be skated.

Again, in addition to this, we have the valuable experience of hosts of enthusiastic and intellectual figure skaters who adopted for many years the radius of 7 feet suggested to them, and of many who have settled down to 6 feet, but who at present have not gained sufficient confidence to enable them to cry "Eureka." It is the search after a truly correct radius that is so fascinating to me; because a learner of curvilinear skating sets himself out to keep his balance and engrave upon the ice according to his powers certain geometrical figures, but with one instrument only. My sympathies are all with him.

I have had very considerable experience of the effects of radii, viz., from 20 feet to 2 feet.

The extreme I fix for combined skating is 9 feet, and

for single figures of the epicycloid variety, such as loops, crosscuts, etc., is 3 feet. If, now, you average the two kinds you get to 6 feet for all-round purposes.

Again, it may be only a coincidence, but it is certainly singular that if you take the average height of man and add to it the distance he is elevated by his boots and high skates from the ice, his radius would not be so very far off 6 feet,

Now, only a certain portion of the entire radius of the blade is in use at a time; this I call the

WORKING RADIUS—W.R.,

and how much this is under certain conditions I will give later on. But what I desire particularly to impress upon the reader is, that its exact location on the blade may vary greatly, and will be governed by the position of the skater's body, whether stooping or upright, as the case may be. It will be fixed by his centre of gravity, for the time being. It is in the skater's power, therefore, to alter the situation of the W.R., and this I regard as a most valuable property.

Again, when the skater is in motion, only one half of the working radius is employed in cutting the groove, and the whole weight will be on this foremost half of the W.R.; the remaining half follows in that part of the groove already cut, and its office then is to assist in guiding.

And now I come to an interesting inquiry, viz., on what portion of this W.R. will turns be made It is clear that the skilful skater will choose the point of least resistance, just before the blade emerges from the groove, according to the direction he is going, whether forwards or backwards. In a system of figure skating this is defined for the purpose of education by the well understood term, "Toe and Heel," not necessarily the very extremes. I think this is right, for the bearing of a blade of 9 feet radius at $\frac{1}{8}$-inch penetration is $10\frac{1}{2}$ inches, nearly the whole length of the blade.

It is a truism to say that turns are more easily made on a blade with a small radius. I became early acquainted with this fact, and as turns have to be made in a forward and backward direction, I determined to follow up the fundamental teaching of toe and heel turns by making another experiment with a skate blade, which I designed with two radii, a fore and aft, at an interval as in the drawing (Fig. 5).

Figure 5: A double radius.

The idea I had was that, as a small radius greatly facilitates turns, the two would answer for forward and backward turns, as each was tilted up, whilst a big curve could be made when the blade rested on the two points in the circumference of the small circles.

I tried this skate thoroughly. The turns, of course, came easily, but wanted care—far too much indeed for practical purposes.

The big curve was too big; in fact, the bearing of the skate at the circumference of the two circles produced really a straight line difficult to alter to a curve. As a whole this skate failed, but it was an excellent instructor and proved incontestably the value of one simple radius.

It is singular, though that after an interval of thirty years a somewhat similar skate blade has been patented in America as a speed skate, the object being to have less bearing. I saw an engraving of this skate in the "English Mechanic" lately.[9]

Skate blade with an adjustable radius

The advantage of having a radius suitable to the individual skater has stimulated the inventive genius of the Swedes.

Mr. Krause, of Gothenburg, has invented a skate blade which is flexible, and by means of a thumb-screw the radius can be increased or diminished within certain limits. There are engravings of it in the "All England Series of Skating," by Douglas Adams, p. 91, Ed. 1890.[10]

I have no knowledge of its success, or other wise, but in my opinion a fixed blade of suitable radius selected after trial would be better, as I consider that moveable parts in a blade in which rigidity is a *sine quâ non* are highly objectionable.

Chapter II. The radius of the blade in an inclined position

At a first glance it might be thought that the parallel-sided blade would not help the curvilinear skater very much, and that he would not only have to cut a groove, but grind off the ice, as it were, by a twisting action to the intended curve.

But it is here that the radius of a right angle blade shews the perfection of the principle. The base lays itself down in the direction of such curve. It may be well seen by magnifying the principle.

This is necessary to properly understand it, because the curve of a skate blade and the curves made by it are so large that the governing principle is somewhat obscure, although existing in the minutest degree.

On turning to Plates IV and V (pages 60 and 61), and holding the photo to the level of the eye and looking at it horizontally it will be seen that the groove represented by the white part is curvilinear, the curve line shows that even in a state of rest this parallel-sided disc does not lay down as a straight line, but as a curve.

Indeed, if I could impart to the radius a personality I should say that the skater is personally conducted over his curves as long as they are of the ordinary nature.

The epicycloid curve, of which the loop is a good example, requires a small radius, and no doubt a torsion action is desirable at particular times.

A very good idea may also be obtained of the nature of this set down by placing any disc of metal of a circular form, such as a coin, at an angle on a soft material, like melted sealing wax, or dough or putty, and then pressing it in slightly. The impression will be curvilinear at the base.

Penetration or depth of the cut in of the blade

This quality is so closely connected with the effects of radius, that its consideration naturally follows in order.

Penetration varies according to the width of the iron, the particular angle at which the blade is sharpened and its actual sharpness, the quality of the ice, the weight of the skater and the existing amount of centrifugal force.

The weight of the skater's body at rest, inclined sideways and supported, will hardly make his skate scratch hard ice, but when in action all this is changed. The skate then in the inclined position cuts into the ice. Now a very important question arises; how deep I have often examined the apparent depth of cut and found it very difficult to ascertain correctly, owing to the presence of the severed ice crystals in a semi-powdered state. On good hard ice it is very small, say $\frac{1}{32}$ or $\frac{1}{16}$ of an inch, on soft ice it may reach $\frac{1}{8}$. Assuming to be the extreme and then considering the effect of radii of 3, 4, 5, 6, 7, 8 and 9 feet conjointly with certain angles, I have been enabled to work out some tables.

The angular position and the other factors employed in their construction

From observation and experiments on the ice, I estimate the extreme angular position of a skater at 25° for single figures of the loop character. For the combined figures I fix the extreme at 15°. Then as a basis, ample for the enquiry, I select angles at the regular intervals of 5°, 10°, 15°, 20° and 25°.

Then these angles are associated with the radii aforesaid, 3, 4, 5, 6, 7, 8 and 9 feet.

Then again angles and radii are associated with penetration of $\frac{1}{32}$, $\frac{1}{16}$, $\frac{3}{32}$, and $\frac{1}{8}$ inch.

Angles, radii and penetration being thus the three potent causes which affect the action of the blade of the figure skate.

Explanation of Plate I

This shows the skating angles, also other angles outside skating requirements, but alluded to on page 73, also the penetration.

Explanation of Plate II

This shows the seven radii under notice. I drew them with beam compasses[11] in the first instance, afterwards I had templets made by an eminent optician and compared them with my curves—they agreed perfectly.

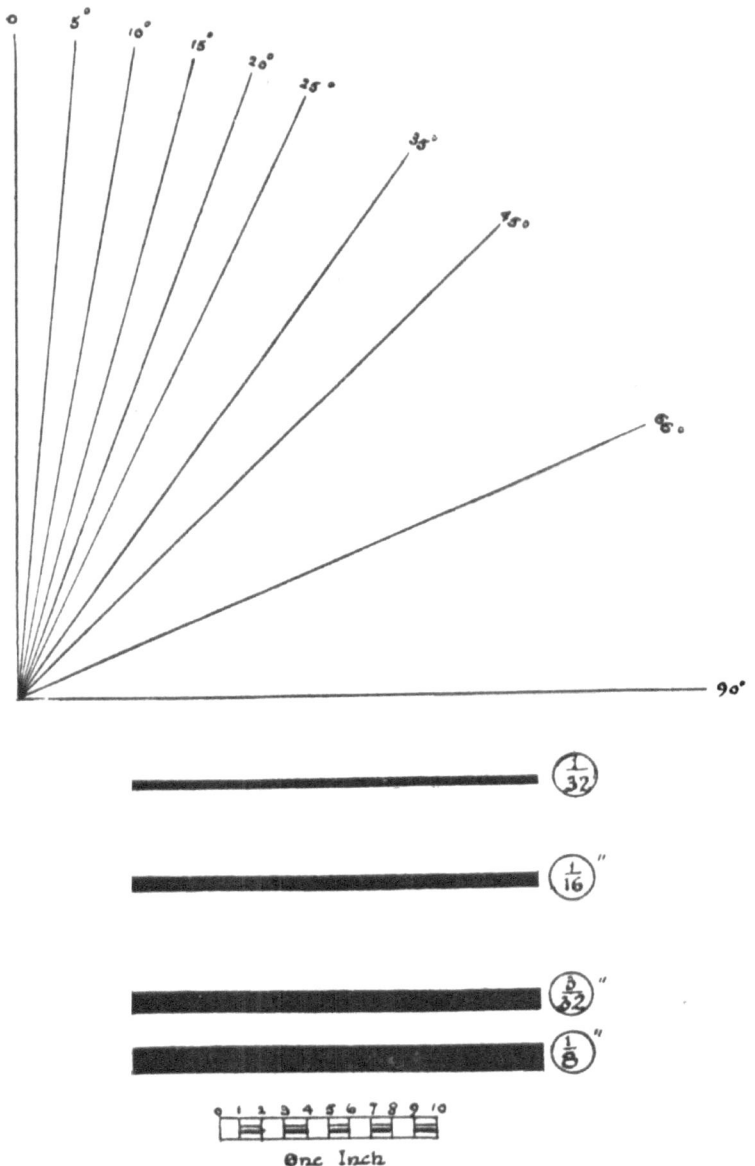

Figure 6: Plate I

The Figure Skate

These radii are best observed by holding the photo horizontally to the level of the eye, when the gradations can be clearly seen.

As they are little more than half the length of a skate blade, they may perhaps appear less curved than an actual skate blade does; this of course is a mere optical effect.

Preliminary investigations and experiments

Before, however, I commenced the construction of the tables I drew the seven radii. Then by the aid of fine pointed dividers and magnifying glass made several measurements of the chords of the respective arcs, at a penetration of $\frac{1}{32}$, $\frac{1}{16}$, $\frac{3}{32}$, and $\frac{1}{8}$, arriving at certain results, which I tabulated for reference.

I also experimented with a measuring apparatus, which I designed and made expressly for the purpose. But I only applied this to measurements of $\frac{1}{16}$ penetration and 6 feet and 9 feet radii. This apparatus was made as follows:—

On a perfectly flat board of mahogany, I veneered a piece of cardboard exactly $\frac{1}{16}$ inch thick, and then cut out a groove upon it $\frac{1}{4}$-inch wide and a segment of a circle of 16 feet. I fitted into this groove two sliding pieces of the same thickness, and placing the skate blade between them, carefully advanced the sliding pieces as far as they would go. Then I removed the blade and measured the distance between the sliding pieces. This I did first in the vertical position, and then having added a

Figure 7: Plate II: Radii

horizontal arm, connected with another moveable arm graduated for angles, I attacked the angular positions, having fixed the blade as wanted.

On the whole this method gave very good results. The experiments were troublesome and numerous, and extended at leisure intervals over a long time, and were conducted with great care. Altogether, I arrived at the conclusion that with a machine of this character, fitted with micrometer adjustments and of Whitworth accuracy, perfect results could be obtained.

Eventually, however, I decided to employ for Table 1, the necessary geometrical problem, and personally to work out all the twenty-eight sums.[12]

Explanation of Table 1

This table relates to the perpendicular position. It is assumed only for the purposes of commencing this inquiry from its foundation that the skate blade is submerged $\frac{1}{32}$, $\frac{1}{16}$, $\frac{3}{32}$, and $\frac{1}{8}$ respectively in very soft level ice, and in the perpendicular position. It then shows in inches and decimals the chord of the arc, or the actual horizontal length of the blade which is beneath the surface. This length of the blade becomes increased afterwards in the angular position and then begins to be actually effective, and constitutes the working radius—W.R.

This table, contrasted with Table 2 for the angular positions, shows how the bearing on the ice is gradually increased as the angles increase, and also how the depth of cut, or penetration, increases the bearing in like manner, and finally how the radius effects it.

Radius	Depth of Cut (W.R., inches)			
(feet)	$\frac{1}{32}$ inch	$\frac{1}{16}$ inch	$\frac{3}{32}$ inch	$\frac{1}{8}$ inch
3	2.96	4.24	5.16	6.00
4	3.46	4.88	6.00	6.92
5	3.86	5.46	6.64	7.74
6	4.24	5.98	7.26	8.40
7	4.56	6.48	7.92	9.16
8	4.88	6.92	8.46	9.80
9	5.20	7.34	8.98	10.38

Table 1: The perpendicular position.

Explanation of Table 2: The angular position

This table consists of one hundred and forty trigonometrical calculations, all of which I have personally worked out on purpose to show the effect upon the skate blade of the seven radii, five angular positions, and four degrees of cut in or penetration comprised therein.[13] Hitherto there has been a great deal of wild conjecture about the space occupied by the skate blade upon the ice. Opinions have been freely expressed and laid down without the foundation of any mathematical basis to give them authority.

Figure skaters will, I trust, find in this table reliable data. They will be able, from the information it affords, to select a skate blade of a suitable radius for combined or single figure skating.

The Figure Skate 45

As before, the results are given in inches and decimals, and the bearing on the ice represents the W.R., or working radius.

Table 2

Radius (feet)	Angle (degrees)	Depth of Cut (W.R., inches)			
		$\frac{1}{32}$ inch	$\frac{1}{16}$ inch	$\frac{3}{32}$ inch	$\frac{1}{8}$ inch
3	5	3.000	4.252	5.212	6.004
	10	3.024	4.274	5.232	6.040
	15	3.058	4.314	5.292	6.098
	20	3.108	4.373	5.374	6.182
	25	3.174	4.452	5.460	6.296
4	5	3.482	4.914	6.000	6.942
	10	3.510	4.944	6.040	6.986
	15	3.550	4.998	6.100	7.046
	20	3.598	5.056	6.186	7.152
	25	3.642	5.168	6.298	7.277

Continued on next page

Table 2 – *Continued from previous page*

Radius (feet)	Angle (degrees)	Depth of Cut (W.R., inches)			
		$\frac{1}{32}$ inch	$\frac{1}{16}$ inch	$\frac{3}{32}$ inch	$\frac{1}{8}$ inch
5	5	3.898	5.496	6.674	7.758
	10	3.930	5.526	6.760	7.826
	15	3.970	5.596	6.822	7.886
	20	3.998	5.656	6.922	7.980
	25	4.062	5.772	7.044	8.130
6	5	4.244	6.008	7.386	8.496
	10	4.274	6.041	7.398	8.548
	15	4.308	6.102	7.476	8.630
	20	4.376	6.186	7.592	8.750
	25	4.452	6.298	7.728	8.912
7	5	4.599	6.492	7.982	9.186
	10	4.616	6.530	7.998	9.232
	15	4.662	6.616	8.070	9.328
	20	4.714	6.686	8.188	9.454
	25	4.808	6.802	8.336	9.624

Continued on next page

Table 2 – *Continued from previous page*

Radius (feet)	Angle (degrees)	Depth of Cut (W.R., inches)			
		$\frac{1}{32}$ inch	$\frac{1}{16}$ inch	$\frac{3}{32}$ inch	$\frac{1}{8}$ inch
8	5	4.910	6.946	8.514	9.814
	10	4.944	6.988	8.550	9.876
	15	5.000	7.050	8.629	9.980
	20	5.054	7.156	8.754	10.102
	25	5.154	7.278	8.914	10.288
9	5	5.226	7.374	9.028	10.406
	10	5.236	7.400	9.070	10.472
	15	5.292	7.482	9.160	10.573
	20	5.374	7.596	9.284	10.718
	25	5.454	7.726	9.406	10.916

Chapter III. Skate blades whose sides are non-parallel

Difficult as this part of my subject undoubtedly is, no research into the forms of skate blades would be complete without a due consideration of those whose sides are convex or concave.[14]

Formerly I was in the constant habit of making notes when anything of interest cropped up in my mind, which happily might facilitate the acquirement of the art of figure skating.

The following extract from some manuscript notes and sketch in the year 1873 relates to these forms, which it will be observed are similar to, but more highly curved than, the existing convex or Dowler skate:

"Extract.—Curve of skates must be made the subject of mathematical investigation, especially as to the effect of inclination sideways upon an tron shaped, as in Fig. 8 (left), as well as curved at the bottom in the usual way. There is much in it I am sure that is interesting. The same may be said of an iron shaped as in Fig. 8 (right).

What effect would this have?"

The result of the inquiry I made shortly after was decidedly unfavourable. I arrived at the conclusion, to which I still adhere, and to which a more extended knowledge of the mathematical principles involved confirms me, viz., that any departure from parallelism, be

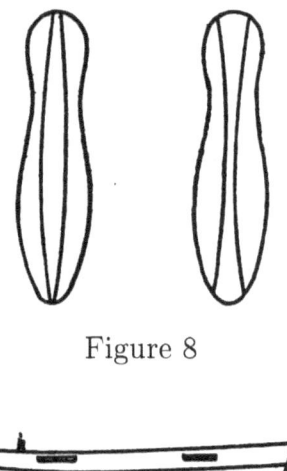

Figure 8

Figure 9: Side view

it convex or concave, is, on account of the elliptical principle evolved, an error, which increases rapidly as either form increases.

Consequently with this conviction I never had a skate blade made either on the convex or concave system. I simply put these forms aside as of no advantage, because I repeat that any attempt to decrease or increase the bearing of a *skate blade abnormally* by means of its side convexity or concavity instead of by the radius of the parallel blade, cannot be done without the introduction of the aforesaid detrimental principle.

It will be at once apparent that the combination of lateral curves with a vertical curve in a skate blade, associated with a varying inclination of the body sideways, produces effects which are very complex, but the consideration of which no figure skater can afford to shirk if he wishes to understand the true principles that govern the skate blade in action.

The convex sided blade

This is a very old pattern indeed; I recollect it and have used it as a boy. It must previously have been in existence a long time; I do not think, however, that such a form was intentional in the first instance. Everyone who has worked at the vice knows well that in filing a surface intended to be flat, the difficulty is to avoid it being convex, and the same remark will apply to material manipulated on the anvil or grindstone. Whether accidental or intentional, this form died out in England, but is much used in foreign countries, consequently its revival here has begun and is increasing.

Never having been patented, however, but as I think quite accidentally adopted, there is no recorded claim as to the nature of the supposed improvement. This is to be regretted, because a specification would have been interesting, as the opposite of that of the concave.

However, this hiatus can be remedied, for the effect of the convex side, coupled with a vertical radius and an inclination sideways, is to decrease the length of the bearing of the blade on the ice, up to a certain angle, but not beyond. When this point is reached the vertical radius overpowers the lateral radius, and an increase of bearing begins; but as this takes place after about 25°, it is beyond the angle of the lean over of the skater.

The convex blade is thought to give increased facilities for turning and pivoting. It is probable that those who think so have never tried a parallel blade of 2 feet radius.

The concave sided blade

In the year 1879, the late Captain Dowler endeavoured to improve the skate blade, and the idea of a concave sided blade also occurred to him, quite independently of mine, but some six years later. He arrived at opposite conclusions to me as to its value to skating, and took out a patent for it. In the specification, dated July 19th, 1879, we get his ideas as to this skate, and I therefore extract those parts of it which are necessary.

"My Invention relates to a construction of the blades of skates in such a manner as to facilitate their use, especially for movements in curves. For this purpose, instead of making the blade with flat or straight sides, as it is usually made, or with convex sides, as it has sometimes been made, I make it with concave sides, so that the blade is thinnest at or about the middle of its length, and widens out both ways towards the ends, the increase of thickness resulting from a gentle curvature to which the concavity is formed. The blade has its flat or running face curved in the ordinary way. By adopting this form of blade when the skate is canted to either side for the purpose of moving in a curve, the edge of the blade having a curvature in the same direction as the curve to be performed by the skater, such curve is per formed more easily than when the blade is straight on its sides or made with a convex curvature. The concavity above described may be only on one side of the blade, the other side being made straight, but it is of advantage to make both sides concave so as to give facility in performing curves in either direction."

The last paragraph is singular. The patentee tolerates one side of the blade being made straight; then goes on subsequently to say—

"According to this formation it is hollowed at the sides so that it is thinnest at or near the middle of its length, and swells out in thickness with a gentle curvature towards the ends. Owing to the vertical curvature at the lower edge of the blade the skate when vertical will rock or tilt lengthwise, and also if it had straight sides, and still more if it had convex sides, it would rock or tilt lengthwise when inclined from the vertical; but by making the sides more or less hollow or concave, according to my Invention, there will be a diminution of such lengthwise rocking or tilting movement when the skate is inclined from the vertical, as for running in curves, that is to say, there will in such attitudes be a greater portion of the length of the blade in actual or approximate contact with the ice, and this is practically found to give increased stability as well as facility in traversing curves. The amount of concavity given to either or both sides of the blade may be varied, but practically I find that a skate works well when each side is hollowed out to the extent of $\frac{1}{12}$th to $\frac{1}{16}$th of an inch, the blade having the vertical curvature usually adopted in skates used for figure skating.

"Having thus described the nature of my Invention and in what manner the same is to be performed, I claim,

"The construction of a skate blade hollowed on either or both sides, substantially as and for the purposes herein set forth."

The concave sided or Dowler skate was launched with considerable skating authority, my friend Mr. Witham giving energetic aid to its introduction. Figure skaters generally were eager to welcome any skate which was supposed to benefit the art of figure skating.[15]

I saw it when first tried at the Skating Club. It then had a lateral radius of 12 feet, and a vertical one of 7 feet. I had a foreknowledge of what it would be likely to do from my previous study of such a form. As the angular inclination increased it scraped the ice and sent out a shower of ice spray when on a curve, besides cutting in very much at the toe.

Afterwards the lateral radius was increased to 16 feet, which diminished the concavity and lessened the bearing on the ice, and in a very great measure corrected the above faults. Then the narrowest part of the curves was shifted further back, and the cutting edge very cleverly modified by Mr. Witham to a right angle, as in its previous state it was too sharp, *vide* "Badminton," page 63, published 1892; article, Figure Skating—T. M. Witham.[16]

In short it is gratifying to have to record that everything has been done to make this skate successful. But at the same time it is most important to observe that all the quasi improvements made since its first inception have been to *decrease* the original bearing on the ice. The convex form apparently has not required any delicate attentions from its admirers.

Writers who have alluded to the concave skate seem to take for granted that the side curve, which is concentric, is the same on the ice. This is not so, however, for when the skate is canted over at an angle the contact of the skate blade side curve with the plane surface of ice is no longer concentric but elliptical.

This skate became very popular. It has been and still is used by thousands, who are doubtless attracted by its form.

A learner wants to make a curve and he sees in this form of blade a curve ready made, as it were, and expects it to help him. In addition to this there is the word "patent." Its influence is powerful.[†]

The concave sided blade is set forth by its advocates as the greatest improvement of the modern figure skate. On the other hand we have the fact that the Figure Skating International Championship was gained by a skater using the convex blade.

Convex and concave blades

Thus we have for consideration these two forms, equally popular amongst their respective admirers. Now if we accept the deduction that a parallel one-edge or V blade

---

[†]NOTE.—In proof of this an order recently came to a well-known skate maker for a pair of speed or racing skates to be fitted with Dowler blades which are essentially figure skates. *Vide* specification.

is perfect in theory, and I fail to see how it can be denied, it follows that no such V blade could be applied to the convex or concave forms, even in theory.

Everyone must admit that the convex and concave are mutually so antagonistic in principle that they cannot by any possibility both be right. Here then we have something definite to start from, and in this respect they are eminently useful to the subject under examination, for by a simple average, without any mathematics, they furnish to any reasonable mind an incontestable proof that the parallel sided blade is the true form, and one more corroboration of the force of the words of Ovid

In medio tutissimus ibis.[17]

But even this will not be sufficient for anyone in either class of believers in the convex or concave respectively who convinced "against his will is of the same opinion still."

I must therefore address myself more particularly to figure skaters, who may be wavering in their opinion as to the best skate to adopt, and ask them to carefully consider the following mathematical facts, so that they can ascertain the merits or demerits of the convex or concave blade. I must confess that to me it seems rather sad that such irreconcileable views should exist about these skate blades. I have known able figure skaters forsake their first love, the parallel blade, take up with the concave for a season, get discontented with it, and then go with a bound right over to the convex.

Surely this unsettled state of mind about the proper

form of an instrument adapted for engraving certain geometrical figures upon the ice is astounding and certainly ought not to exist.

The first proposition is that the convex sided blade, say of 6 feet vertical radius, 14 feet lateral radius, and $\frac{1}{16}$-inch penetration, continually decreases its bearing on the ice when inclined at an angle up to about 25°, but from that point it begins to increase. The concave, on the other hand, goes on increasing its bearing, but not beyond 66°, as after that it begins to form an arch.

The second proposition is that the guiding principle of the blade formed by its contact with and penetration into the ice at an angle is elliptical owing to the convexity or concavity employed.

This is really the root of the matter, as the guide of an ellipse if it is actually used to any appreciable extent, is bound to scrape or else cut off more than is wanted; consequently the skater must be in such a case struggling against it.

If I can prove that the convex and concave contain this grave error, then I shall have to ascertain how it is that these forms so much in use are practicable at all. This, however, is very easy.

Following out my previous idea to make these matters more intelligible, I find that it is necessary to magnify the principle which influences them. In doing so I shall treat the three forms—the convex sided, the parallel sided, and the concave sided—together. Plate III represents photos of three models I made of wood and turned up accurately in the lathe. The side curves of the convex and concave are exactly the same, $2\frac{1}{2}$ inches

Figure 10: Plate III: Models of curves.

radius, all three are $3\frac{3}{8}$ inches long and have a uniform vertical radius of 5 inches. At an angle of 25° in the concave model perfect contact of the entire length of the concave side with a flat surface and on the level is obtained, and this angle of 25° shall be applied to each. Here it might be as well to observe that in the Dowler skate blade of 14 feet lateral and 6 feet vertical radius, with a penetration of $\frac{1}{16}$, perfect contact occurs at 66°.

The three models were then each placed in separate small boxes, and on a level surface, and each model within was, by means of adjusting screws, inclined to the right 25°. I then took casts of them in plaster, then planed down the surfaces, so that the greatest depression in each case was precisely alike. This was $\frac{1}{7}$-inch. The following photos were then obtained:—

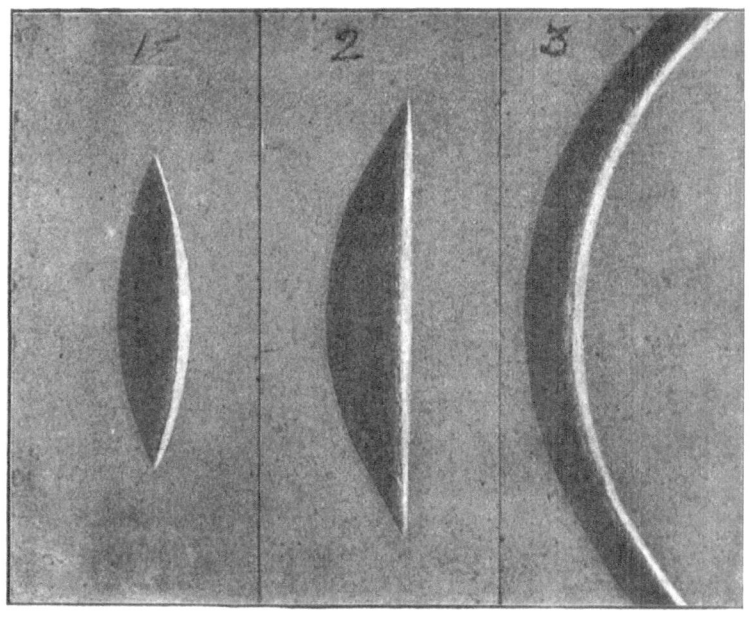

Figure 11: Plate IV

The Figure Skate

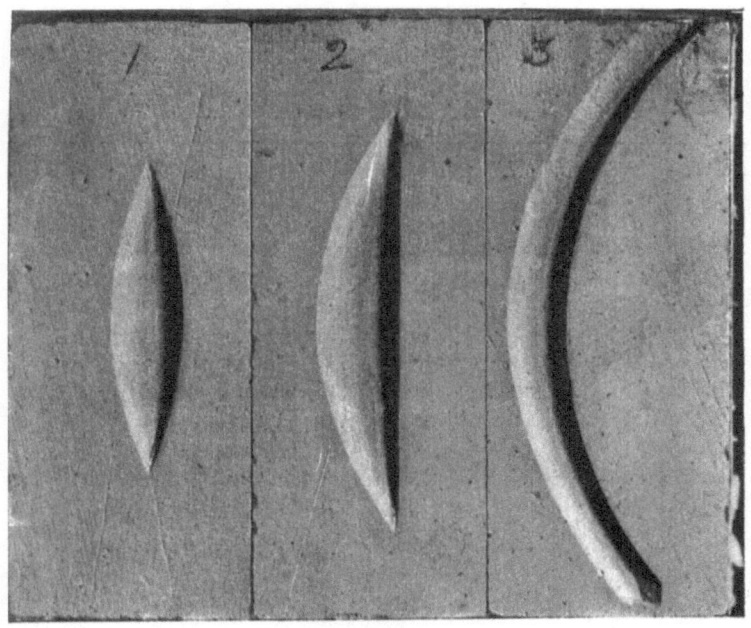

Figure 12: Plate V

Two of these are identical, but they are differently illuminated. In one case the inclined base of the three forms is taken light and the cut dark. In the other the inclined base of the three forms is taken dark and the cut light. By studying these photos together the contour of the cut in each case can be well understood.

It is the impression made by the contact of the lateral and vertical lines at an angle that governs the subject.

On examining that of the convex form (viz.: 1 in Plates IV and V), the startling fact appears that the groove actually points and guides to a curve the very opposite of the one required.

If this principle is really rendered effective in a skate blade, there can be necessarily nothing but a scrape. Besides pointing the wrong way, the curve is elliptical.

The concave groove of No. 3 shows complete contact, the curve being also elliptical as can be plainly seen. The dotted line in Fig. 13 shows the true circle, the plain line the ellipse, and the space between them its extent.

Figure 13

In the photo No. 2, the parallel sided model, shows that the cutting edge is ready to pare off whatever curve is required, neither more nor less, and on very close observation it will be found that the base line at the bottom on the right, or the left on the top of the groove, is curvilinear.

The ellipse thus governs the set down of the convex and concave forms, but not so the parallel, which withstands this severe test and remains concentric. And if it is so in these models, which magnify, it follows that it must be present in actual skate blades; and indeed, I have detected it in the blade under notice. How then are the convex and concave blades at all practical? The answer to this question is that the elliptical principle is so largely diluted that as at present used it is almost harmless. The reader will understand this later on.

Chapter IV. Measurements

I have made a great variety of measurements with my apparatus, and extract a few:—

Concave radius, vertical 6 feet and lateral 14 feet, $\frac{1}{16}$th penetration.

Angle	15°	bearing	6.5
"	25°	"	7.2
"	35°	"	7.75
"	45°	"	8.5
"	63°	"	contact.

After the last, the blade rests on its extreme ends and forms an arch.

But the convex, like the parallel blade, can be inclined to 90°.

In case of any doubt being thrown upon measurements, which from the extreme difficulty of conducting them might reasonably exist, I have determined to resort again to trigonometry to solve once and for all this question.

The branch of that accurate science that has to be employed, viz., calculations of angles in which an ellipse is concerned, is beyond my mathematical knowledge, and even if it were otherwise, I should still prefer to have these very difficult calculations from an undeniable source.[18] I have therefore called to my aid Professor Graham of the Technical College, who has kindly interested himself in this matter.

I do not insert the formula by which this gentleman has arrived at his results, as it would occupy a page or two and probably not interest the general reader. But if any figure skater cares to make similar calculations I can promise him a brain exercise of a high character.

Professor Graham says that the problem becomes very complicated when lateral and vertical curves and angles exist, and that for himself, were he a skater he should, without hesitation, select the parallel blade as the best instrument for our purposes.

Now to give tables for the convex and concave blades on the basis of those I have constructed for the parallel blade would involve the calculation of two hundred and eighty very difficult sums.

No object would be served by this laborious work, otherwise I would gladly undertake it.

It could not advance the practice of figure skating one iota, for the simple reason that the advocates of these forms hold irreconcileable views. Curiously enough, on one subject they are in unison, viz.: that the parallel blade is altogether wrong. The third table, then, will be short but effective. It will deal only with those varieties of skate blades having the popular 6 feet vertical and 14 feet lateral radii, with a penetration of $\frac{1}{16}$ and angles of 15° and 25°. But in addition to these skating angles, it will, for the sake of affording ample information as to what takes place when the angles are extended beyond skating requirements, give in each case the angles of 35°, 45° and 66°. I shall also in this table insert the parallel, so that the three forms

Table 3

Angle	Convex	Parallel	Concave
15°	5.782	6.102	6.429
25°	5.754	6.298	7.355
35°	5.815	6.628	7.923
45°	5.969	7.144	9.439
66°	6.716	9.42	entire contact.

can be contrasted. The calculations of the convex and concave are by Professor Graham; those of the parallel by myself.

If the reader will refer to measurements, page 65, he will find approximate results in certain cases of the concave, which is very satisfactory.

Having now for the first time obtained reliable data, we can leave the regions of conjecture, which have been so long occupied by skaters, and proceed to examine what the convex and concave really do as contrasted with themselves and with the parallel blade.

First I take the angle of 15°, because it is the average of my Table of Angles; secondly, because it will give a very fair chance to the concave blade, to show what it can do to elongate the contact.

But let the reader bear in mind that as far as combined skating is concerned, the angles in use would in all probability not exceed 5° or 10°.

Contrasting the convex and concave at 15° with a penetration of $\frac{1}{16}$, there is a difference between them of .647, and contrasting either form with the parallel, a difference of .320, so that the concave, with all this angle of 15° only lengthens the bearing $\frac{3}{10}$ths of an inch. What it does at 10° is about $\frac{1}{5}$th, and at 5°, $\frac{1}{10}$th.

This insignificant result answers the remarks on pages 57–59, and accounts for a skater having used a Dowler on one foot and a parallel on the other without detecting any difference, and I have no doubt that he could use a convex and concave on either foot.

As to the convex at 25° contrasted with the concave at 25°. Here it is evident that the convex is much better for loops, etc., having a shorter bearing than the concave of no less than 1.601.

A parallel skate blade of 5 feet would be the equal of the 6 feet convex, and its superior in involving no elliptic principle.

A parallel blade of $6\frac{5}{8}$ feet radius would give a longer bearing than the 6 feet concave without any elliptic principle.

It has been supposed hitherto by its advocates that a concave 6 or 7 feet vertical would be the equal of a parallel 9 or 10 feet vertical. My calculations show that a 9 feet parallel has more than 1 inch longer bearing than a 6 feet concave.

Beyond our skating angles we first come in Table 3 to 35°. We find here a difference between the convex and concave of 2.108, and at 45° 3.470, and lastly we come to 66°; the convex is at 6.716, the parallel at 9.42, and the concave entirely in contact.

The calculations of the convex proceed in a curious and somewhat erratic way. The reader will observe that 25° has a less bearing than 15°, but beyond 25° to 35°,

The Figure Skate

45° and 66° the bearing lengthens out again. This is because the vertical radius begins to overpower, as it were, the lateral radius.

Now let us for a moment take a flight of fancy and do in imagination what we cannot do in reality, eliminate the injurious elliptic principle and then consider the convex and concave without it.

It is evident that the convex would be a much better skate for all-round purposes than the concave, because the more the angle becomes increased the less bearing is required, and this is what the convex does. The concave, how ever, acts in the opposite way—it increases the bearing where it is not wanted.

From imagination we return to reality; the elliptic principle will not be barred. It is true that you may use it in a very highly diluted sense, as in the instance under notice, with little injurious effect, but certainly with no advantage. But directly you attempt to make it actually effective it rapidly deteriorates. As to a better stroke with the concave there remains the fact that a skater using the narrow-pointed convex wins the championship, and the speed skater gets an impulse equivalent to 20 miles an hour from his parallel blade.

Finally the best, I fear, that can be said of these two forms of skate blades is that they have created a certain stimulus and called forth the imaginative powers of the users of them. Figure skaters who have read the foregoing pages are in a position to form their own conclusions. As for myself I have no other object in view than to arrive at the truth of the matter.

Therefore I can write without prejudice, but with

the courage of my opinion, and I can say that I would eagerly welcome any real improvement in skate blades to the advantage of figure skating, but I can really see nothing in the convex or concave, after this searching inquiry, to dethrone the parallel from its well-deserved position, but on the contrary everything to enhance its value as far and away the best form of blade for curvilinear skating.

Therefore I shall now leave the vagaries of the convex and concave forms, and return to the parallel blade, and shall endeavour to find out what radius will best suit the skater, according to the purpose intended.

To do this effectually it is absolutely necessary to divide the skaters into two classes, first those who are very agile, and have a natural aptitude to rotate the body, and secondly those who have not the latter gift but require help from the blade.

Having done this, then we must also divide the skating movements into combined and single. We take as radii 3, 4, 5, 6, 7, 8 and 9 feet. First as to combined skating, the specially gifted can use 9 feet easily and naturally. The non-gifted will have to impart a considerable amount of forced rotation on his axis in certain movements if he wishes to use this grand blade.

If he cannot do this to the required extent, he must fall back on the 8, 7, or even 6 feet, until he gets suited.

Now for single figures, especially in such a test figure as inside and outside continuous loops. A radius of 3 feet will be far the best to begin with, afterwards using 4, 5 or 6 if possible. For all-round skating a 6 feet will be as efficient as possible under such circumstances.

The Figure Skate

I am convinced that it is the aptitude to rotate that should govern the selection of a large radius for combined skating, as distinct from the small radius that is absolutely required for single figures, especially when these embrace loops, cross-cuts, and all very "curly work," and it is in these selections that Table 2 will be found most useful. Whatever bearing is wanted it can be worked out.

It is only by the exercise of considerable skill that 9 feet can be made to yield inside and outside loops, and this it does unwillingly; whereas, with the 3 feet, with a less bearing of 3 inches, this class of work proceeds with facility.

Generally, I think, skaters will find from a study of Table 2 that there is a much longer bearing than they have hitherto supposed.

Penetration affects this also in a marked degree; for instance, there is double the bearing between $\frac{1}{32}$ and $\frac{1}{8}$ penetration, 9 feet 25°, or roughly no less than $5\frac{1}{2}$ inches.

Consequently the skater travelling over ice of different density, and using a variety of angles, will find the blade of his skate continually varying its length of bearing; defying calculation.

What a contrast to this is the wonderful simplicity of the roller skate, with only its four surface contacts and its most ingenious bogey platform, enabling the angular inclination to adjust more or less correctly the radial action of the rollers. The width of each separate roller, however, being a surface contact, causes the outer and inner edges of the roller to travel on a curve at a different velocity, as they are not the same distance from the centre of such curve, so that there must be a slight skidding action, how ever imperceptible it may seem. Convex or concave rollers are unknown.

It is probable that from the conditions involved there can never be an absolutely perfect ice skate for all-round figure skating. Large and small curves require a different radius.

If competent figure skaters who have leisure time would furnish themselves with blades of the seven radii I have selected, and go thoroughly to work on each, making notes of the effects produced and the kind of ice used, it would be a useful work. It would either confirm existing opinions as to what should be the best radius for an all-round figure skate, which necessarily must be a compromise, or it would set up a new standard.

Such a standard would probably be discovered by progressions of $\frac{1}{2}$-inch added to existing radii.

Thus we might find that a $6\frac{1}{2}$ feet was better than a 6 or 7 feet, or that a $5\frac{1}{2}$ excelled a 6 feet. The difference of bearing at 15° with of penetration, as given by Table 2 (an average of 6 or 7 feet), is .257 inches either plus or minus.[19]

It is fortunate that there evidently is a wide latitude before efficiency is greatly disturbed, but it is much to be desired that the very best radius of such efficiency should be arrived at by working on the lines I have indicated.

The ideal ice skate no doubt would be one in which angular inclination could automatically alter the radius of the parallel blade.

A digression—Thoughts on the possibility of describing the hypocycloid curve by means of turns

At the outset permit me to say that experimenting in the later period of my skating days, I worked hard to accomplish this curve, but utterly failed; hence the name I give it, the Paradox.

It may be asked, "Why then do you bring it forward?" I do so for the following reasons: In the first place it is extremely interesting and fascinating. In the next I confess that I should very much like to have it thoroughly tried by modern up-to-date figure skaters. If eventually it should really be found impossible, why there is an end to my chimera. But if by any subtleties of our art it could be made possible a good work would be done for figure skating.

It is a very different thing to depict mathematical figures on the drawing board, and try to reproduce them on the ice.

In the latter case, one prominent factor is ever present as an obstacle; the difficulty of balancing in novel and trying positions.

Having made these preliminary remarks, I proceed to my task.

The turns made by the figure skater are replete with scientific interest, the outcome of them being illustra-

1. Epicycloid & epitrochoid

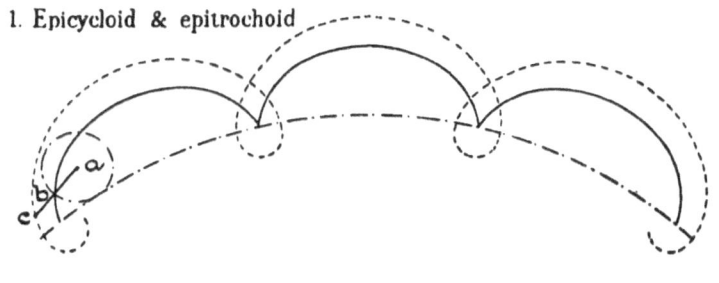

2. Cycloid & trochoid

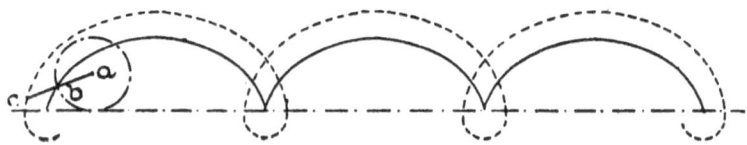

3. Hypocycloid & hypotrochoid

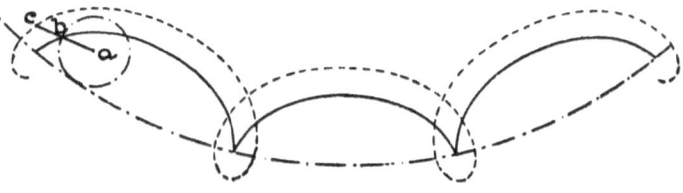

Figure 14: Plate VI

tions of the epicycloid and cycloid curves, and variations of them (*vide* the chapter on Loops, etc., "A System of Figure Skating").[20]

What the nature of these wonderful curves is, and how they have exercised the minds of the most profound mathematicians of ancient and modern times, is a matter of history.

It is rather awe-inspiring for the modest figure skater to find himself endeavouring to engrave some of them upon the ice. But in his attempts he is brought face to face with these curves *nolens volens*.[21] Therefore, in my opinion, his mental enjoyment will be increased and his actions much facilitated, by carefully considering (if he has not already done so) the genesis of the epicycloid, cycloid and hypocycloid curves. Especially will this be the case if he is going to attempt the latter.

On page 74 these three curves are depicted, accompanied by the loops belonging to them,

In the diagram which I have had specially drawn for me by an expert, the plain line represents the curves, the dotted line the loops, and the chain line the base, or as I shall in future call it for my special purpose the contour. The small circle A may represent the skater and B or C the skate.

The first that claims attention is

The epicycloid

In this the contour is curved and the small circle A, the skater, rolls on the outside of such curve, the point B or C represents the skate, the result being the epicycloid and epitrochoid respectively.

These curves are abundantly used in figure skating; they give us the well known Figure[22] 3 and its family and corresponding loops and cross cuts. Without these figure skating would be a very contracted art indeed.

The cycloid

Here the contour is straight, the same rolling action goes on; A represents the skater, B and C the skate—the result is the cycloid and trochoid.

These are so much more difficult than the epicycloid that they are rarely seen in figure skating.

The hypocycloid

In this case the contour is curved, but in an opposite direction to the epicycloid.

The rolling action takes place on the inside of the contour. A represents the skater, B and C the skate, the result being the hypocycloid and hypotrochoid respectively.

It will be observed, on careful examination of the diagrams, that the curves of the hypocycloid are flatter and the loops longer and narrower than those of the epicycloid.

It is very important to consider this, as it must affect the nature of the turn. The whole operation which is required seems simple enough, and the curves of the hypocycloid appear to be those of "relative equilibrium." But I failed to get the contour curved as in the diagram. May others be more fortunate.

The paradox

The way to attempt this figure is to start with the intention of making, say two turns and three curves (none less will show it), but to endeavour to make the contour of the group a straight line, instead of curved as in the epicycloid.

If this is done cycloid curves will be the result. This may be called the first stage, and no further progress can possibly be made until it is attained.

The first stage being accomplished, and proceeding on the same principles, the skater must now start again, and endeavour to bend or curve the contour of the group of turns past and beyond that of the cycloid, until such contour becomes outwards, and thus exactly the reverse of the epicycloid.

If this can be done the paradox will be resolved.

I regret that I can give no further practical information beyond the above, but I can suggest that if this fails it would be desirable to draw the curves on the ice with any suitable pigment, and then endeavour to follow the lines. These lines, or bands, might be half an inch wide so that they could be readily seen.

I adopted a similar method once in practising a very large loop.

28, ALDRIDGE ROAD VILLAS
WESTBOURNE PARK, LONDON

The Author will be glad to hear of your experience or success with THE PARADOX.[23]

Commentary

Notes

1. Here is the relevant excerpt from T. Maxwell Witham, *A System of Figure-Skating: Being the Theory and Practice of the Art as Developed in England, with a Glance at Its Origin and History*, fifth ed. (London: Horace Cox, 1897), 11-13:

> The city of Oxford developed a school of skating of its own many years ago, which seems to have been in advance of what was then generally known in figure skating, and the dexterity attained was owing to the improved form of the skate-iron which was invented by a resident in Oxford, Mr. Henry Boswell. The skating, Mr. Boswell says, consisted of outside edges backwards and forwards, threes single and double, and *loops!* The great difficulty with which these figures were skated on the skates then in use, caused Mr. Boswell to turn his attention to the form of the skate-iron, and he and one or two other enthusiasts at Oxford experimented with irons of different lengths and different curves, and the final result of their numerous trials was to make a skate-iron without the projecting toe, and with the heel elongated and rounded; making it in fact the same shape as the modern club skate.
>
> When by means of the most careful ex-

periments the form of the iron was finally determined upon (and the results of Boswell's experiments demonstrated to him that a curve having a radius of about seven feet was the best), he took the pattern to a Birmingham maker, and had some four dozen pairs of irons made, and these were supplied to the skaters in Oxford. With this new form of skate, skating made rapid progress, and in the year 1838 a club was formed, the members being principally mechanics, tradesmen, and College servants, and skating in combination was started.

The figures consisted of circles and parts of circles, of outside and inside edges, strung round a common centre with variations by means of the cross roll and threes. By the kind permission of Mr. Henry Warner, a pupil of Boswell, I reproduce some of these figures in the ladies' chapter, as they are simple of execution and very effective.

It is curious to note that in America and elsewhere, where combined skating is in its infancy, the first attempts at combination consist of the same movements as are set forth in the archives of the Oxford society.

The great aim of the members of the Oxford society in combined skating was accuracy, and the attention they paid to this accounts in a great measure for the dexterity obtained. The demand for Boswell's skates was so great that the making of them was taken up by a Sheffield firm, and the improved form of iron came into general use. Mr. Boswell was not only the ingenious inventor of the skate, which, so far as the iron is concerned, is the club skate of the present day, but he was also considered the best skater in the Oxford society, and, with one of the other members, was professionally engaged to skate on the artificial ice which had been moved from Baker-street to the Coliseum.

2. The Skating Club was a prestigious group of figure skaters in London.

3. Compare this type of skate with the klapskates currently used by speed skaters. The concept is the same.

4. My Dowler skates attach to the boot using the Mount Charles system. It features sole and heel plates that screw to the boot and clamp the blade. See figure C5 on page 90.

5. Fowler dedicated his little book to Vandervell. It has been republished, with extensive commentary, as G. Herbert Fowler, *On the Outside Edge: Being Diversions in the History of Skating*, ed. B. A. Thurber (Evanston, IL: Skating History Press, 2018).

6. At about 0.15 inches thick, modern figure skate blades

intended for freestyle or general skating are slightly thicker than Vandervell recommends. Dance blades, at 0.11 inches thick, are slightly thinner.

7. Long-time skaters may remember the Pattern 88 blades manufactured by John Wilson, which had holders that attached to the boots and interchangeable blades for figures and freestyle.

8. "Degrees of frost" refers to the number of degrees below freezing. "Three degrees of frost" is 29°F.

9. Filor's skate is described in US Patent 617,694 (January 10, 1899). The article Vandervell was referring to is "Filor's Improved Skates," *English Mechanic and World of Science* 49, no. 1769 (February 1899): 10. Here's the text and the image:

Filor's Improved Skates

In the accompanying illustration we present a skate which is provided with an improved lock for the heel and sole clamps, the lock being so constructed that either of the clamps may receive an initial or broad adjustment without disturbing the adjustment of the other clamp. Fig. C1.1 is a perspective view of the skate, and Fig. C1.2 is a top plan view in which the clamps are shown in closed position by positive lines, and in open position by dotted lines. The runner, in order to offer as little friction as possible, is arched, so that it bears on the ice only beneath the heel and sole-plates. Upon the heel-plate a clamp is held to slide, and upon the sole-plate diverging clamps pivoted together at their rear-ends are held to slide. Between the sole and heel-clamps a lever is fulcrumed. Oppositely-curved links pivoted to the lever, one at each side of the fulcrum, are adjustably connected by means of screw-shanks with the sole and heel-clamps. The pivotal connections between the links and the lever are out of alignment with the centre of the fulcrum, so that when the curved portions of the links are brought close together by the movement of the lever, they will lock themselves in this position. As each link is independently adjustable upon its clamp, the throw of the clamps may be separately regulated to change the locking action of the lever. The skate is the invention of Charles P. Filor, care of S. S. Moore, Trenton, N.J.—*Scientific American.*

10. Douglas Adams, *Skating* (London: George Bell and Sons, 1894), 90–91 describes the Krause skate as follows:

> In the Krause skate, the blade of the skate consists of two parts, viz. a top or ridged part, A, and a bottom or sliding blade, B, which are connected at the ends by means of small bolts or pints, CC.

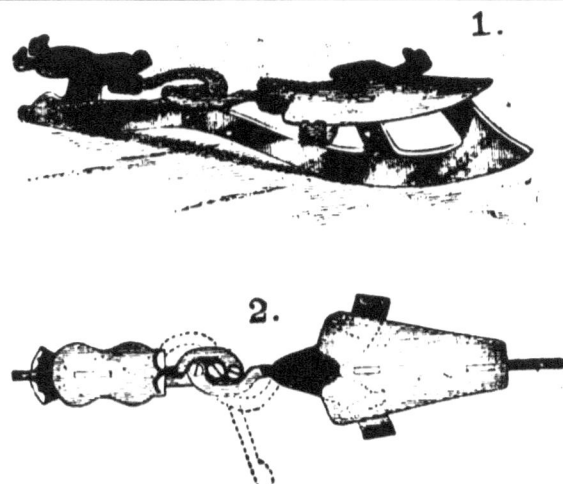

Figure C1: Filor's improved skates.

In the diagram [Figure C2] the sliding blade, B, is shown, for example, furnished with vertical projections, D and E, which pass through slots in the ridged part, A. The front projection, D, of the sliding blade is extended, so that it forms a support for the front or sole-plate, F. To prevent any strain between the sliding blade and the sole-plate, the hole through which the connecting pin is passed is made oblong. At about the middle of the ridged part, A, a screw, G, is applied, which, when it is screwed down, presses against the sliding blade, so that the latter can be curved to any required radius. The bottom end of the screw enters a hole in the sliding blade, which is there to prevent it from moving laterally.

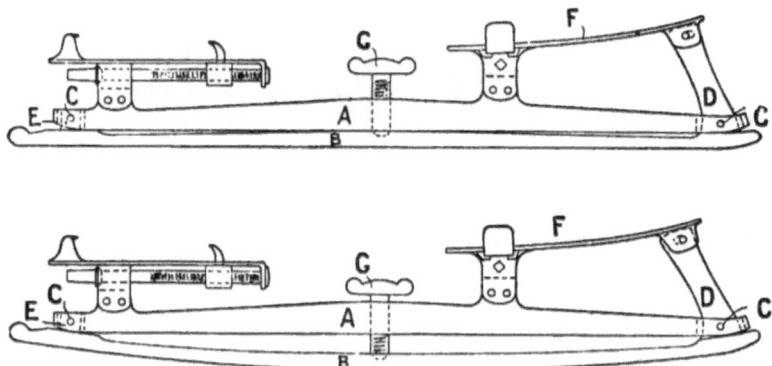

Figure C2: The Krause skate.

11. A beam compass is a large compass that can produce circles with radii several feet long. Scribes, for those who remember using them as an aid for compulsory figures, are beam compasses.

12. Here's how to work out these numbers. In figure C3, the horizontal line represents the ice surface, and the circle represents the skate blade. Of course, a real skate blade is not a full circle, just part of one. The other lines have no physical significance and are drawn to show the values used in the calculations. R is the rocker radius, h is the penetration depth, and L is the length of the skate blade in the ice.

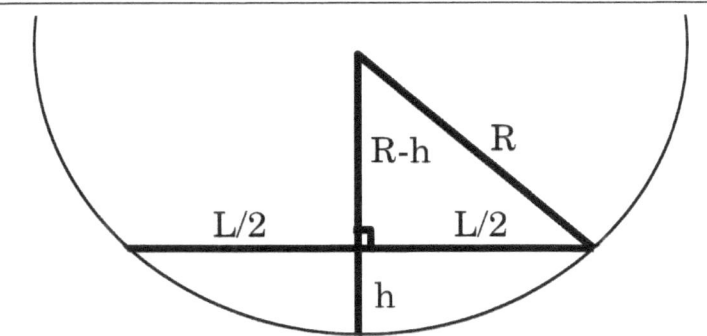

Figure C3: The geometry for the calculations in Table 1.

Table 1 shows the value of L for different values of R and h. L can be calculated using the Pythagorean theorem,

$$R^2 = \left(\frac{L}{2}\right)^2 + (R-h)^2.$$

Solving for L yields

$$L = 2\sqrt{2Rh(1-h)}.$$

Vandervell's values are quite close to the values I found using a computer—generally within 1%.

13. The calculations in Table 2 are a bit more complex than those in Table 1, but not too difficult. The geometry is shown in figure C4. The difference is that now, the blade—represented by the gray line—is tilted at angle θ relative to the vertical. The horizontal line

is still the ice surface, but now we're looking from the back of the blade instead of from the side. d is the penetration depth (the value given in the table), and h is the amount of blade in the ice.

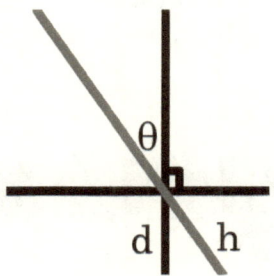

Figure C4: The geometry for the calculations in Table 2.

This geometry leads us to the conclusion that $h = d \sec \theta$, which can then be plugged into the formula used in Table 1,
$$L = 2\sqrt{2Rh(1-h)}.$$
With our new h, this becomes
$$L = 2\sqrt{2Rd \sec \theta (1 - d \sec \theta)}.$$

Once again, Vandervell's calculations agree well with mine.

14. Over a century later, some skate manufacturers are still experimenting with concave blades. Examples include the "parabolic" blades produced by John Wilson,

which are thinner in the middle than at the ends. Their concavity is not nearly as pronounced as that of the Dowler skate in figure C5!

15. The Dowler skate is rather alarming. Its blade flares substantially out at both ends, as shown in figure C5.

Figure C5: The Dowler skates in the editor's collection. The bottom skate shows the flaring at the ends of the blade.

16. Witham's ("Figure-Skating," 62–63) comments on the angled skate blades are as follows:

> Formerly, everyone used acute-angled edges. Around the year 1875 all the members of the Skating Club went to the other extreme, and had their skates ground to obtuse-angled edges; but that lasted only a

few weeks, after which they tried right angled edges, and this happy medium seems now to hold its own. In Fig. C6 illustrations are given of acute, obtuse, and right-angled edges.

Figure C6

I dismiss the acute angle as quite unsuitable for figure-skating; it is so sharp that it sets up great friction by cutting in too deeply, and yet almost all the common sort of skates are ground to an acute-angled edge.

The obtuse-angled blade errs in the opposite direction. Used by a heavy man on soft ice, it is delightful, but if the ice is hard there is a very unpleasant feeling of insecurity attached to its use. At the same time there is no doubt that its form minimises friction, and its blunt edge enables difficult turns to be accomplished with less danger of 'catching' the ice in making the turn; but even its most ardent admirers admit that, for the first ten minutes or so, if on hard ice, they experience a feeling of insecurity which is induced by the blunt edge. Personally, I always skate with right-angled edges and take care to have them sharp, as I believe that the confidence which sharp edges give, and the extra power of striking off, more than compensate for the little extra friction which is consequent on the cutting in of the sharp edge. The right-angled edge is a happy medium between the acute and the obtuse, and is the form of edge generally used by good figure-skaters. The great object to be attained in constructing a skate-blade is to make it of such a form as will reduce friction to a minimum. A skate having a curvature of nine-feet radius would do this in consequence of the distribution of the skater's weight over a large bearing surface, but then turns and loops would become very difficult. Obtuse-angled edges are used for the same reason, as although the bearing surface is not increased, the blunt edge cuts in very little, and consequently sets up but little friction. The Swedes and Norwegians, who in the last few years have taken to figure-skating of an acrobatic character, have their skates not only ground to a five-foot radius, but increase the pivoting power by having the sides made convex.

17. "In medio tutissimus ibis" means "on the middle course you will go most safely." However, Vandervell has misquoted Ovid. As Terry notes in "In medio sta-

tio mediocria firma locantur," *Notes and Queries. Sixth Series* 9, no. 215 (February 1884): 135, Ovid's phrase is "medio tutissimus ibis."

18. The calculations for convex and concave blades are indeed much more difficult.

19. Vandervell calculated this value by taking averaging the two values selected from Table 2 (6.616 and 6.102), averaging then, and dividing the result by two:

$$0.257 = \frac{6.616 + 6.102}{4}.$$

20. Here is the relevant section of Witham, *A System of Figure-Skating: Being the Theory and Practice of the Art as Developed in England, with a Glance at Its Origin and History*, 143–146:

> Before attempting to explain to my readers the nature of Loops and Crosscuts and the practical way of skating these figures on ice, I wish to lay before them the diagrams and definition of the three varieties of the cycloid and trochoid curves.
>
> If a circle EPF roll along a straight line A B (C7) so that every point of the circumference may touch the line in succession, and if P be that point of the circumference which was in contact with the straight line at the beginning of the motion, when the

circle has made a complete revolution the point P will have described a curved line A P D B, which is called a cycloid or trochoid.

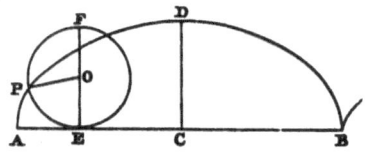

Figure C7

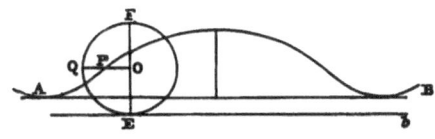

Figure C8

Again, if a circle roll along a straight line E b (Figs. C8 and C9), and Q be that point of its circumference which was in contact with the straight line at the beginning of the motion, and P be a given point in O Q, the radius of the circle (Fig. C8), or in the radius produced (Fig. C9), when the circle has completed a revolution the point P will have described a curved line at P D B which is called a prolate or inflected cycloid or trochoid if the point be within the circle (Fig. C8), and is called a curtate cycloid or trochoid if the point is without the circle.

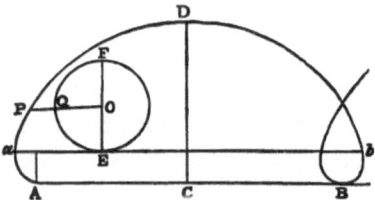

Figure C9

The curves or figures used by the skater may be grouped under two heads, the simple and the compound, of the cycloid and epicycloid families; and from close observation of the effect they respectively have upon him, I consider that they mark two distinct styles of skating, and for the following reasons.

Upon the Simple Figures, which are parts of circles, by far the best style and attitude are attainable, and the skating of the expert becomes large, bold, flowing and graceful, because the body can be kept erect and quiescent.

Upon the Compound Figures (of which the ordinary loop is an example, and which is known as the curtate cycloid), the additional rotary motion necessarily given to the body, and the peculiar position often demanded of it, sadly detract from the style; and the movements of the skater are, to a certain extent, ungraceful, because he will often (especially if he be a tall man) have to lower his centre of gravity by bending his knees and stooping; the unemployed leg, too, so useful as a means of imparting rotation, will naturally desire to fly out even to the horizontal position on account of the centrifugal force.

Numberless aerial illustrations of the wonderful curves, cycloid, epicycloid, &c, &c, may easily be obtained by suspending a bullet from a string held in the hand, and then making it into a rotating pendulum, at the same time moving the point of suspension in any required direction.

The cycloid may vary in form according to the nature of the line or base along which the generating circle or skater revolves, and also as to the position of the point tracing that circle, whether it is on, within, or beyond its circumference, and assuming the base line to be straight in the first instance, it will produce a common cycloid (which if continued two revolutions answers to the curve of Carré which is part of the cardioid of Castilliani, known to skaters as the common 3); in the second case a prolate or inflected cycloid; and in the third, a curtate cycloid (the ordinary loop). When the base line is curved we pass to the epicycloid curves, which have interior or exterior cusps, and the epitrochoid, which have interior or exterior loops. Loops on the ice may thus vary in form, at one time flattened, balloon or pear-shaped, at another, the lines may come nearly to a point or cusp, at another actually cross, and in this instance when further carried out we have "the crosscut," to be hereafter described, a most remarkable figure. The wonderful properties of the cycloid or "curve of quickest descent" account for the skater appearing for a moment stationary, and then actually bounding forward again, and it is at this remarkable pause that, if he so desire, he can execute the "crosscut," which is performed by suddenly shifting the base line some eight, twelve, or more inches, backwards or forwards as the case may be.

When a skater is describing a cycloid curve in a given direction, say for instance forwards, if he suddenly alter his direction as mentioned above, this does not alter the rotation he started with, consequently that portion of the movement which is in reverse, may be a straight or curved line according to the *modus operandi*, and the speed at which the reverse is introduced.

I desire here to call attention to the prolate or inflected cycloid, which doubtless has much to do with curious curves which occur in some of our operations. It will be observed that as the describing point is within the circle as it rolls along its base, curves counter to one another as in the serpentine lines, are produced. Is it possible to skate a figure of this character? The execution by the skater of cycloid curves is extremely fascinating, but it appears almost a physical impossibility to perform them with elegance, for the reasons I have previously given: but admitting that after due attention to its defects a few accomplished and powerful skaters may be able, by disguising the rotation of the body and by keeping the unemployed leg under control, to preserve some trace of the better attitude belonging to the simpler curves, still I feel it my duty to raise a warning voice to all, and particularly young learners of skating, who have not graduated in the simpler curves, to beware how they follow the gambols of the enchanting sirens of the ice, trochoid curves, which will otherwise most certainly beckon them, down "the curve of quickest descent," into the regions of bad style, from which there may be no escape.

Much more fascinating material is to be found in the chapter; in fact, the entire book is quite interesting.

21. *Nolens volens* is Latin for "unwilling or willing."

22. The figure 3 is known as a three turn today.

23. As the editor of this book, I took it upon myself to attempt this challenge. Since I can no longer write to Mr. Vandervell, I'll describe my experience here. I found it relatively unproblematic to do three-turns halfway around the inside of a pre-drawn circle. I kept them tight, with the edges between no more than a couple of feet long, alternating right forward inside and right back outside. Right forward outside loops were a little more troublesome, but quite doable. I think Vandervell's difficulty may stem from his insistence on using the English style, which is designed for large, sweeping curves rather than the tight turns that make skating the hypocycloid practical.

Further reading

There are not many books that get into the nuts and bolts of skate design and how it affects skating. Here are some that I've stumbled across.

A direct follow-up to Vandervell's study is C. S. d'Este Stock, *The Figure Skate: A Research into Dimensions and Their Effects Upon Performance with a Consideration of Penetrations into the Ice and Pressure upon It* (Folkestone, Kent: A. Stace and Sons, 1954), which even has a strikingly similar title. It's available in the University of Connecticut's online collection of skating books at `https://archives.lib.uconn.edu/`. Inspired by Vandervell's work, d'Este Stock performs similar experiments with different blades. He concludes "that designed skates are better than the commercial article and that improvement would be neither difficult nor expensive" and ends with a prediction:

> I believe that the artistic skating of the future will make use of the undeveloped possibilities of variation in design of blades and that detachable blades, so successful in our own skate, and the best for ensuring a high degree of precision in manufacture will be used. They will be made for "ad hoc" movements. I imagine a series of blades, easily

changed, for special movements which can then be well executed by skaters of quite moderate skill.[‡]

Nothing could be further from what actually evolved.

Those wishing to know more about skate design and the details of sharpening can do no better than to read Sidney Broadbent, *Skateology: A Technical Manual for Skaters Regarding Skates, Skating Fundamentals, Skate Sharpeners*, Revised (Littleton, CO: ICEskate Conditioning Equipment, 1997). This self-published manual can be ordered at http://www.iceskateology.com/.

[‡]d'Este Stock, *The Figure Skate: A Research into Dimensions and Their Effects Upon Performance with a Consideration of Penetrations into the Ice and Pressure upon It*, 31–32.

Bibliography

Adams, Douglas. *Skating*. London: George Bell and Sons, 1894.

Broadbent, Sidney. *Skateology: A Technical Manual for Skaters Regarding Skates, Skating Fundamentals, Skate Sharpeners*. Revised. Littleton, CO: ICEskate Conditioning Equipment, 1997.

d'Este Stock, C. S. *The Figure Skate: A Research into Dimensions and Their Effects Upon Performance with a Consideration of Penetrations into the Ice and Pressure upon It*. Folkestone, Kent: A. Stace and Sons, 1954.

Duguid, Charles. *The Story of the Stock Exchange: Its History and Position*. London: Grant Richards, 1901.

"Filor's Improved Skates." *English Mechanic and World of Science* 49, no. 1769 (February 1899): 10.

Filor, Charles F. *Ice-Skate*. US Patent 617,649, filed September 23, 1898, and issued January 10, 1899.

Fowler, G. Herbert. *On the Outside Edge: Being Diversions in the History of Skating*. Edited by B. A. Thurber. Evanston, IL: Skating History Press, 2018.

Hines, James R. "Vandervell, Henry Eugene (1824–1908)." In *Historical Dictionary of Figure Skating*, 233. Plymouth, UK: Scarecrow Press, 2011.

Terry, F. C. Birkbeck. "In medio statio mediocria firma locantur." *Notes and Queries. Sixth Series* 9, no. 215 (February 1884): 135.

Vandervell, H. E. *The Figure Skate: A Research into the Form of Blade Best Adapted to Curvilinear Skating.* London: Straker Brothers, 1901.

Vandervell, H. E., and T. Maxwell Witham. *A System of Figure-Skating: Being the Theory and Practice of the Art as Developed in England, with a Glance at Its Origin and History.* London: Horace Cox, 1869.

Witham, T. Maxwell. *A System of Figure-Skating: Being the Theory and Practice of the Art as Developed in England, with a Glance at Its Origin and History.* Fifth ed. London: Horace Cox, 1897.

———. "Figure-Skating." In *Skating*, 41–197. Badminton Library of Sports and Pastimes. London: Longmans, Green, 1892.

Illustration credits

Cover Skater from H. E. Vandervell and T. Maxwell Witham, *A System of Figure-Skating: Being the Theory and Practice of the Art as Developed in England, with a Glance at Its Origin and History* (London: Horace Cox, 1869), Plate 1. Table of numbers from page 39 of the copy of H. E. Vandervell, *The Figure Skate: A Research into the Form of Blade Best Adapted to Curvilinear Skating* (London: Straker Brothers, 1901) in the collection of Harvard University Library. Both were digitized by Google Books. The top drawing and back cover illustration are figures 4 and 14, respectively, in the present volume; see below.

1–2 From the copy of H. E. Vandervell, *The Figure Skate: A Research into the Form of Blade Best Adapted to Curvilinear Skating* (London: Straker Brothers, 1901) in the British Library (General Reference Collection 07905.g.38). ©The British Library Board.

3–14 From the copy of H. E. Vandervell, *The Figure Skate: A Research into the Form of Blade Best Adapted to Curvilinear Skating* (London: Straker Brothers, 1901) in the collection of Harvard University Library, digitized by Google Books.

C1 From "Filor's Improved Skates," *English Mechanic and World of Science* 49, no. 1769 (February 1899): 10, digitized by Google Books.

C2 From Douglas Adams, *Skating* (London: George Bell and Sons, 1894), digitized by Google Books.

C3–C4 Drawn by B. A. Thurber.

C5 Photograph by B. A. Thurber.

C6 From the copy of T. Maxwell Witham, "Figure-Skating," in *Skating*, Badminton Library of Sports and Pastimes (London: Longmans, Green, 1892), 41–197, in the collection of B. A. Thurber, digitized by B. A. Thurber.

C7–C9 From the copy of T. Maxwell Witham, *A System of Figure-Skating: Being the Theory and Practice of the Art as Developed in England, with a Glance at Its Origin and History*, fifth ed. (London: Horace Cox, 1897) in the collection of Harvard University Library, digitized by Google Books.

Also available from
Skating History Press
Publishing new editions of historic books about skating.

A Treatise on Skating
R. Jones
First published in 1772, this is the oldest surviving book about ice skating. This edition includes the full text of the original work, the sheet music for "The Skater's March," and W. E. Cormack's 1855 revisions, plus a new introduction and notes.
98+iv pages, illustrated

On the Outside Edge
G. Herbert Fowler
A clever and insightful history of skating, this book traces the development of figure skating through the late nineteenth century. This edition—the first since 1897—uses the results of current scholarship to bring Fowler's work up to date.
145+iv pages, illustrated

Frostiana
A souvenir of the last great Frost Fair, *Frostiana* has a reputation for having been printed on the frozen Thames. It contains numerous amusing anecdotes about winter events and activities. This edition includes a new introduction and period illustrations.

165+vi pages, illustrated

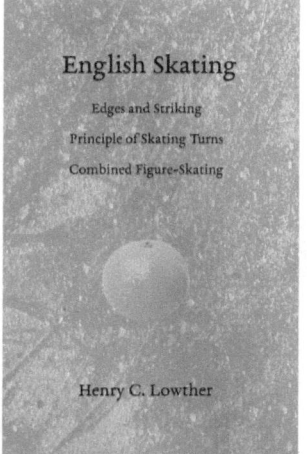

English Skating
Henry C. Lowther
Edges and Striking, Principle of Skating Turns, and *Combined Figure-Skating,* brought together in this volume, provide a comprehensive picture of English skating at the end of the last century and give today's skaters tips on improving their skills.

227+iv pages, illustrated

http://www.skatinghistorypress.com/

www.ingramcontent.com/pod-product-compliance
Lightning Source LLC
Chambersburg PA
CBHW030156100526
44592CB00009B/308